About Demos

Who we are

Demos is the think tank for everyday democracy. We believe everyone should be able to make personal choices in their daily lives that contribute to the common good. Our aim is to put this democratic idea into practice by working with organisations in ways that make them more effective and legitimate.

What we work on

We focus on six areas: public services; science and technology; cities and public space; people and communities; arts and culture; and global security.

Who we work with

Our partners include policy-makers, companies, public service providers and social entrepreneurs. Demos is not linked to any party but we work with politicians across political divides. Our international network – which extends across Eastern Europe, Scandinavia, Australia, Brazil, India and China – provides a global perspective and enables us to work across borders.

How we work

Demos knows the importance of learning from experience. We test and improve our ideas in practice by working with people who can make change happen. Our collaborative approach means that our partners share in the creation and ownership of new ideas.

What we offer

We analyse social and political change, which we connect to innovation and learning in organisations. We help our partners show thought leadership and respond to emerging policy challenges.

How we communicate

As an independent voice, we can create debates that lead to real change. We use the media, public events, workshops and publications to communicate our ideas. All our books can be downloaded free from the Demos website.

www.demos.co.uk

First published in 2005
© Demos
Some rights reserved – see copyright licence for details

ISBN 1 84180 151 8
Copy edited by Julie Pickard
Typeset and produced by Land & Unwin, Towcester
Printed in the United Kingdom

For further information and
subscription details please contact:

Demos
Magdalen House
136 Tooley Street
London SE1 2TU

telephone: 0845 458 5949
email: hello@demos.co.uk
web: www.demos.co.uk

The Public Value of Science

Or how to ensure that science really matters

James Wilsdon
Brian Wynne
Jack Stilgoe

DEMOS

DEM©S

Contents

Acknowledgements   7

About the authors   9

Foreword   11

1.   The scientific state we're in   15

2.   Lost in translation   24

3.   Upstream – without a paddle?   30

4.   Citizen scientists   45

5.   Challenges great and small   51

     Notes   61

Acknowledgements

Last September, Demos published *See-through Science*, which made the case for 'upstream' public engagement in science and technology. We have been encouraged by the discussion that this has generated, and are grateful to those who have provided comments or critique.

In developing this new version of our argument, we have talked to a number of scientists, social scientists and policy-makers, whose insights have helped to advance our thinking. These include Barbara Adam, Darren Bhattachary, Peter Cotgreave, Claire Craig, Rob Doubleday, Ted Freer, Steven Gentleman, Robin Grove-White, Richard Harvey, Sheila Jasanoff, Richard Jones, Gary Kass, Matt Kearnes, David King, Ragnar Lofstedt, Phil Macnaghten, Claudia Neubauer, Steve Rayner, Dan Sarewitz, Simon Schaffer, Alister Scott, Isabelle Stengers, Andy Stirling, Bronislaw Szerszynski, Jim Thomas, Alexis Vlandas, Johannes Vogel, James Warner, Claire Waterton, Mark Welland, Becky Willis, Robert Winston and the staff of the Alzheimer's Society.

This pamphlet is an initial output of the NanoDialogues project, which is partly funded by the Office of Science and Technology's Sciencewise programme (www.sciencewise.org.uk). We are grateful to our partners in that project: Biotechnology and Biological Sciences Research Council (BBSRC), Engineering and Physical Sciences Research Council (EPSRC), Environment Agency, Lancaster University and Practical Action. Particular thanks to members of the steering group for their guidance: Rachel Bishop, Adrian Butt, David Grimshaw, Phil Irving, Elizabeth Mitchell, Chris Snary, Gillian Thomas, Richard Wilson and Monica Winstanley.

Brian Wynne's contribution forms part of his responsibility for the theme of public engagement with science in the ESRC Centre for the Economic and Social Aspects of Genomics at Lancaster University.

The pamphlet's launch has been organised in partnership with the British Association for the Advancement of Science, and we are grateful to Roland Jackson and Sue Hordijenko at the BA for their advice and support. Finally, at Demos, our thanks to Tom Bentley, Molly Webb, Paul Miller, Julia Huber and Sam Hinton-Smith for their contributions at various stages. None of these organisations or individuals necessarily endorses all our views, and any errors remain our own.

James Wilsdon
Brian Wynne
Jack Stilgoe
September 2005

About the authors

James Wilsdon is Head of Science and Innovation at Demos. He advises a wide range of organisations on science policy, emerging technologies, democracy and sustainability, and his recent publications include *See-through Science: Why public engagement needs to move upstream* (with Rebecca Willis, Demos, 2004); *Masters of the Universe: Science, politics and the new space race* (with Melissa Mean, Demos, 2004); *The Adaptive State: Strategies for personalising the public realm* (edited with Tom Bentley, Demos, 2003); and *Digital Futures: Living in a networked world* (Earthscan, 2001). He is an Honorary Research Fellow at Lancaster University and an Associate Director of Forum for the Future (james.wilsdon@demos.co.uk).

Brian Wynne is Professor of Science Studies at Lancaster University and a senior partner in the ESRC Centre for the Economic and Social Aspects of Genomics (CESAGen). His research has focused on the relations between expert and lay knowledge in science policy, innovation, environment and risk assessment. He was a special adviser to the House of Lords inquiry into Science and Society in 2000, and a member of the Royal Society's Committee on Science in Society (2001–2004). He is the author or editor of numerous publications, including *Misunderstanding Science? The public reconstruction of science and technology* (with Alan Irwin, CUP, 1996), and *Science and Citizens: Globalisation and the challenge of engagement* (with Ian Scoones and Melissa Leach, Zed Books, 2005) (b.wynne@lancaster.ac.uk).

Jack Stilgoe joined Demos as a researcher in January 2005. He works on a range of science and innovation projects. Previously, he was a research fellow in the Science and Technology Studies department at University College, London, where he looked at debates between scientists and the public about the possible health risks of mobile phones. He has a degree in economics, an MSc in science policy and a PhD in the sociology of science. He has recently published papers in the journals *Science and Public Policy* and *Public Understanding of Science* (jack.stilgoe@demos.co.uk).

Foreword

Robert Winston

Our lives are increasingly affected by scientific discovery and the application of science has undoubtedly brought huge benefit to society. People live healthier, longer lives in greater material comfort and few, in any country, would choose to live without the benefits of technology. Some people claim that technical sophistication has reduced our appreciation of the simple, fulfilled life. Perhaps this is true to some extent. But happiness is a difficult state to measure, and it is probable that, throughout the world, technological advances have increased the measure of human happiness.

There are negative aspects that we have been too ready to ignore. Whilst people have always greedily grasped at technological advance, it has been perceived by others as a threat to human wellbeing. This was as true in ancient Rome at the time of Pliny, as it was during the industrial revolution in nineteenth century Britain. Whether scientists like it or not, technological advance is now increasingly seen as a massive threat – to mankind, and to our planet. And rightly so. Recent history is not reassuring. We remember vividly the unchecked power that some military weapons have. And it is simply not enough for some distinguished scientists to assert that the undoubted dangers of technology – for example, global warming, nuclear disaster, genetically enhanced 'humans' – can be easily controlled by different technology, or better regulation by governments.

People generally are well-informed and discerning and it should not astonish us that so many view the science we value with suspicion – or even hostility. It should not surprise us that this suspicion is most acute amongst people living in the developed world, from whence much of our advanced technology emanates. Even in long-established democracies, people do not feel that they have ownership, control or even much influence over the technologies that are exploited by their governments and by commercial enterprises.

The scientific community once believed it could assuage public concerns over the misuse of science by better communication of the benefits of scientific knowledge. There has been gradual, sometimes grudging, recognition that mere communication – whilst important – cannot alleviate justifiable anxieties. Now the watchword is 'engagement' and with it, 'dialogue'. The scientific community is beginning to realise, but often reluctantly accept, that we scientists need to take greater notice of public concerns, and relate and react to them. Expressions of despair at public ignorance, impotent polemics about the advantages of technology, assertions that our economy is threatened by reactionary attitudes, attempts at manipulation of the press, are all totally inadequate responses. Neither will mere lip-service about the value of public engagement be helpful.

Hence it is a privilege to write a foreword to this pamphlet. Demos has taken various steps to explore the issue of public engagement, and this publication is the latest contribution. Of course, many inside and outside the scientific community will not agree with all the conclusions tentatively implied here. For example, so-called 'upstream engagement', where members of a concerned public recommend what research might be most useful, may be legitimate for a charitable organisation like the Alzheimer's Society whose donors have a narrow objective. But it may be less relevant for publicly accountable bodies like the Medical Research Council that are responsible for health research on much broader fronts. The Research Assessment Exercise, which this pamphlet considers has 'few friends', may still be the most appropriate of a number of somewhat inadequate mechanisms for distributing the government support that is so vital to continued excellence in our universities. And the authors of this document may be right to express concern at the increasing commercialisation of university research, though by no means all will agree. It undoubtedly produces many benefits, but it can limit academic freedom, and may sometimes encourage pursuit of applied research not in the best interests of basic scientific enquiry.

The time is right for examining the means and the details of public engagement. One step forward might be for the scientific community to accept that it does not own the science that it pursues. Another step may be for government to place more value on proper public dialogue, and to facilitate it better. This pamphlet is a valuable contribution to this vitally important debate.

Lord Winston is Professor of Fertility Studies at Imperial College London and Director of NHS Research and Development for The Hammersmith Hospitals Trust. He was the former Chairman of the House of Lords Select Committee on Science and Technology and is the current President of the British Association for the Advancement of Science. He regularly presents BBC science programmes, including most recently 'Human Instinct', 'The Human Mind' and 'Child of our Time'.

1. The scientific state we're in

In the autumn of 1985, the morale of British scientists hit an all-time low. Budgets were being slashed, student numbers were in freefall, and there was a steady drain of talented minds to the US. Most worrying of all, despite Margaret Thatcher's own background as a chemist, science remained low on her list of priorities. The Royal Society sent regular delegations to meet Sir Keith Joseph, then Minister for Education and Science, but were told that, regrettably, Britain simply couldn't afford to spend more.

A small group of academics decided to take matters into their own hands. They circulated a letter to friends and colleagues, and within a few weeks had secured donations and expressions of support from 1500 scientists and engineers, including 100 Fellows of the Royal Society and several Nobel Prize winners. They used this money to take out a half-page advert in *The Times*. Under the headline 'Save British Science', it warned that 'Whole areas of research are in jeopardy. . . . There is no excuse: rescue requires a rise in expenditure of only about one percent of Government's annual revenue from North Sea oil. We can and must afford basic research.'[1]

Twenty years on, the picture looks remarkably different. In March 2005, the lobby group Save British Science – created as a result of that original *Times* advert – renamed itself the Campaign for Science and Engineering. Its director, Peter Cotgreave, explains why they decided to strike a more positive note: 'We needed to reflect the way the political climate for science has been transformed. Research budgets are as high as they've ever been and are still rising. A lot of scientists, especially the younger ones, kept asking me "What do we need saving from?"'[2]

There is no denying that science and innovation occupy a privileged place in New Labour's agenda. Since 1997, the science budget has more than doubled, and Gordon Brown has set out a ten-year framework for future investment, with the aim of boosting R&D to 2.5 per cent of national income by 2014.[3] In 2002, Tony Blair became the first prime minister to deliver a speech at the Royal Society, and announced his ambition to make the UK 'one of the best places in the world to do science'.[4] Science was mentioned 11 times by Labour in its 2005 election manifesto (compared with five by the Lib Dems and none at all by the Conservatives). Peter Cotgreave argues that this is one area where the government displays a real unity of purpose: 'Blair is fascinated with modernity, and for him science is all about the future. Brown has done a hard-headed analysis and worked out that if we're going to pay the bills in ten or 20 years' time, we need to be doing more science now.'[5]

A crisis averted?

As this extra money starts to trickle through the research system, some feel they have a further reason for optimism. The once frosty relationship between science and society appears to be thawing. Five years ago, the House of Lords identified a 'crisis of confidence' and called for 'more and better dialogue'.[6] But as the public and media debate over genetically modified (GM) crops intensified, many scientists felt they were on the losing side of a battle for hearts and minds. There was a consolation prize: the GM saga made the scientific establishment sit up and think about the importance of dialogue on difficult issues. As a result, they have adopted new, and better, models of science communication. There is a growing confidence that lessons have been learned.

Two pieces of evidence are commonly cited in support of this view. First, the debate over nanotechnologies, which at one stage threatened to snowball into a GM-style controversy, is now being held up as a model of scientific self-regulation and early public engagement. The inquiry by the Royal Society and Royal Academy of Engineering, and the government response that followed, are seen as a successful template for managing the ethical, social and environmental dilemmas posed by emerging technologies.[7] Second, there are signs that public opinion is starting to swing back in science's favour. The latest MORI poll commissioned by the Office of Science and Technology (OST) shows that 86 per cent of people think science 'makes a good contribution to society' – up 5 per cent on two years ago.[8]

No one tries to deny that serious mistakes were made in the past. But BSE and GM are increasingly spoken of as aberrant blips, rather than episodes which highlight deeper, more systemic problems in the governance of science and technology. There is a creeping sense of complacency within some sections of government and the scientific community: a belief that we can return to business as usual, with a few new committees and a little extra public consultation, but without any fundamental reform of scientific culture and practice. As one policy-maker describes it:

> At a rhetorical level, the language has moved on. Look at [Lord] Sainsbury's speeches in the past year. Something has shifted. But if you look beneath this, the issues are being dealt with in a very half-hearted and scattergun way. It just isn't a strategic priority. Which leading scientists are really driving this?[9]

Hitting the notes but not playing the music

A contrasting view is that the real work has just begun. The House of Lords report called for reflection and dialogue on social and ethical issues to become a 'normal and integral part' of the scientific process.[10] On this account, while there has undoubtedly been progress, there is still a lack of clarity as to how new approaches can be embedded in the policies, practices and institutions of science. And in some quarters there remains downright opposition, stoked up by critics of reform.[11]

The British Association for the Advancement of Science (BA) Festival of Science is a good barometer of this lingering sense of unease. Each September, the BA President opens the Festival with a speech on a science and society theme. Tensions and uncertainties have dominated in recent years. In 2002, Sir Howard Newby admitted that 'the public now feels that it is reduced to the role of a hapless bystander or, at best, the recipient of scientific advance and technological innovation which the scientific community believes it ought to want.'[12] In 2003, Sir Peter Williams warned that 'scientific dark ages' lie ahead if the public fails to recognise 'the overwhelming benefits science brings, rather than always tending to look to the dark side'.[13] And, last year, Dame Julia Higgins argued that 'only by entering a real dialogue, admitting the risks as well as hailing the potential benefits of new knowledge, will we maintain the respect and trust of society, and restore it where it has been damaged'.[14]

Beneath the thin crust of consensus in these debates there lies a deeper ambivalence. Old assumptions continually reassert themselves. To give one recent example, Alec Broers, in his 2005 Reith Lectures, *The Triumph of Technology*, rehearsed the now familiar argument that 'it is time . . . to move away from the old concept of "the public understanding of science" to a new more dynamic "public engagement"'.[15] Minutes later, in the debate that followed, he had this exchange with Mary Warnock:

> Baroness Warnock: *After the election, the government, whatever government, has simply got to bite the bullet and start planning and constructing new nuclear reactors. In spite of your extremely welcome insistence that the public must be involved, do you think the public is really well-enough informed? Are they not perhaps too apprehensive to make this decision? It seems to me that what is needed here is very firm leadership.*

> Lord Broers: *I agree with you. But I don't know how quickly we can educate the public, to bring the evidence forward in a calm and rational way.*

No sooner have 'deficit' models of the public been discarded than they reappear.[16] Similarly, there are persistent misunderstandings of what is at stake in these discussions. Participation tends to be invited on narrow 'risk' questions when, as we argued in See-through Science, the public are equally concerned about the wider social visions and values that are driving science and innovation.[17] And some still fear that listening to the public will act as a 'brake' on scientific advances, rather than a source of improved social intelligence and better decision-making.[18]

Science, technology and everyday democracy
In this pamphlet, we argue that despite the progress that has been achieved over the last five years, a fresh injection of energy and momentum is now required. Otherwise, we will end up with little more than the scientific equivalent of corporate social responsibility: a well-meaning, professionalised and busy field, propelled along by its own conferences and reports, but never quite impinging on fundamental practices, assumptions and cultures.

Today's hard work needs to becomes tomorrow's routine. How do we reach a situation where scientific 'excellence' is automatically taken to include reflection and wider engagement on social and ethical dimensions? Such expectations cannot be externally imposed. If they are to take root, they must be nurtured by scientists and engineers themselves.

Success will also require a new humility on the part of those non-governmental organisations (NGOs) and social scientists who advocate reform. For understandable reasons, many have concentrated on the 'hardware' of engagement – the methods, the focus groups, the citizens' juries – that can give the public a voice in science policy and decision-making. But in the next phase, this needs to be accompanied by a greater appreciation of the 'software' – the codes, values and norms that govern scientific practice, but which are far harder to access and change. These prevail not only within science, but also around it, in funding and policy worlds. Steve Rayner, who leads the Economic and Social Research Council's 'Science in Society' programme, recognises this challenge:

> *There's a problem in the way social scientists have positioned themselves in these debates. . . . Our constant call for more public participation sidesteps wider issues of responsibility and culture. There isn't a crisis of trust in science, there's a crisis of governance.*

It is tempting to see this as a rather arcane issue to be tackled by the small clique of 'science and society' sympathisers. But we believe its implications are far-reaching. Why is it so important to get it right? Despite the flood of new money, there are still reasons for concern about the long-term future of British science.

First, there are the unpredictable effects of globalisation. In what Thomas Friedman has dubbed the 'flat world', countries such as China, India and South Korea are investing enthusiastically in their science base.[19] A recent study by ScienceWatch found that Asia is rapidly catching up with Europe and the US in the volume and quality of its scientific publications.[20] At the same time, a gradual process of 'offshore innovation' is getting underway, as corporate R&D starts to flow to the new hubs of Shanghai, Seoul and Bangalore. There is a growing political concern that these emerging 'science powers' will undermine Britain's strengths in higher-value research, just as they have already in manufacturing and back-office services, such as call centres. Few ministerial speeches on the economy are now complete without a reference to the 120,000 computer science graduates that China and India are producing each year. On returning from a trip to China in February 2005, Gordon Brown admitted that 'what I have seen has opened my eyes to the sheer scale of the challenge'. And he reaffirmed: 'we have got to seize the China challenge. . . . the next phase of our government's programme must show that we can become world leaders in science.'[21]

A second problem is the lack of new blood flowing into British science and engineering. Student numbers may be booming in Asia, but here they continue to decline. Between 1991 and 2003, despite an overall increase in the number of A-level entries of around 7 per cent, there were falls of 18.7 per cent in chemistry, 25.4 per cent in mathematics and 29.6 per cent in physics. Only biology has bucked the trend. This has had serious repercussions for the university sector, with Exeter's chemistry department the most high-profile of several closures in the past year. A recent editorial in *Nature*, while applauding the substantial improvements in funding since 1997, argued that: 'the scandal of the Blair government's record on science is to be found in the universities. . . . Behind it all lies a lack of joined-up government in addressing the supply and demand for future researchers.'[22]

A third issue is the relationship between science and business. All of this extra funding is sharply conditional: science must deliver economic success. To meet the targets in its ten-year framework, the government is relying on business to match its extra spending on R&D. So far, there is little sign of this happening. In 2003, business R&D rose by 2 per cent – in line with inflation, but far short of the 5 or 6 per cent year-on-year increases that the ten-year framework requires. To help fill the gap, government is pumping more money into supporting collaborative research between business and universities.[23] Yet this policy response, while entirely sensible, is indirectly fuelling a different set of concerns. There is growing disquiet among university scientists that the drive for ever closer ties with business – enthusiastically promoted by the Lambert Review[24] – is distorting research priorities. At its most benign, the prospect of corporate funding may tempt researchers away from high risk, novel areas of research towards more readily marketable applications. But some fear that it has more subtle effects on the integrity of the research system. In a recent *Guardian* article, the neuroscientist Steven Rose lamented that:

> *I have never felt so seriously competitive. . . . As patenting has become so common, as industry has moved into the campuses, it is competition, not cooperation, which is at a premium. Even within the same lab, there can be Chinese walls between researchers funded by different sponsors. We no longer speak openly about our most recent work at scientific conferences, because to do so would give our colleague-competitors a head start.*[25]

Our aim in this pamphlet is to explore how these challenges relate back to the science–society relationship: how the big economic and political dilemmas facing science are inextricably bound up with social questions of trust, governance, democracy and public value.

We describe some of the efforts that scientists and engineers are making to open their work to new influences, and outline some experiments that are under way to encourage ongoing reflection and engagement with the wider world. But we start, in the next chapter, by asking if these tasks could be made easier with a shared framework for describing, debating and building the 'public value of science'. Drawing on related discussions in public services and the arts, we explore whether the concept of public value could be applied to science and technology as a way of raising questions that often remain unasked: What is scientific knowledge for? How should we imagine its purposes, futures and meanings? Is producing 'wealth' the only criterion? Or are there other measures of public value?

In chapter 3, we revisit the notion of 'upstream' public engagement, and respond to some of the questions and criticisms that the idea has attracted. We discuss the limitations of linear models of innovation, and acknowledge the implied linearity in our own metaphor of the stream. Rather than seeing public engagement as a one-off fix, we emphasise the need for engagement throughout the complex and varied stages of innovation. We then give examples of organisations that are pioneering new approaches, including the Alzheimer's Society and the Medical Research Council.

Chapter 4 focuses on the role that 'citizen scientists' can themselves play in creating more reflective cultures of science, and the way in which the norms and values that govern science are evolving to reflect new forms of social knowledge and accountability. Lastly, chapter 5 outlines some specific challenges that scientists, policy-makers and others need to address in order to move this agenda forward.

Our focus throughout is on publicly funded science, although we recognise that the lines between public and private are increasingly blurred. And while we tend to use the terms 'science' and 'scientists' as shorthand, we hope that our argument will be read as equally relevant to technologists and engineers, whose contribution will be vital if the public value of science is to be realised.

In all of this, our goal is to identify how science, technology and engineering can be strengthened by what Demos calls 'everyday democracy' – the way we govern ourselves through the choices, commitments and connections of daily life.[26] Scientists need more frequent opportunities to talk about the choices they are making, the assumptions their work reproduces, and the purposes to which it might be directed. Whether it is the prospect of a new generation of nuclear power stations, the convergence between nano and bio-technologies, or novel forms of human enhancement, our capacity for innovation will continue to present us with dilemmas as well as opportunities. But it is our belief that Britain's hope of becoming 'the best place in the world to do science' rests as much on giving scientists and engineers the freedom and incentive to renew their institutions and practices as it does on ten-year frameworks and R&D targets. Developing a more substantial and authentic debate on these questions is in the best interests of science, and of an enlightened democracy.

2. Lost in translation

'Good prose', George Orwell once remarked, 'is like a window pane.'[27] But words, like windows, don't always give us a clear view on the world. They may be smudged, tinted or curved in ways that distort what we can see.[28] In a perceptive study of the language used during the GM controversy, Guy Cook, a professor of applied linguistics, draws our attention to the role that words play in defending particular interpretations of GM technology and in marginalising opponents. He forensically analyses the language used by different players in the debate, and shows how words and phrases such as 'Luddite', 'sound science', 'Frankenstein foods' and 'tampering with nature' are used to create a particular impression. The different actors talk and write as though their language were indeed a transparent glass, through which the truth about GM can be observed. Yet often, the language used, 'while purporting to be rational, honest, informative, democratic and clear, is in fact . . . illogical, obscure, patronising and one-sided, populated with false analogies, misleading metaphors and impenetrable ambiguities.'[29]

GM might be an acute case, but it is by no means the only example of this problem. Scientists, engineers, social scientists, politicians, NGOs and wider publics tend to think and talk about science and technology in different ways, so that shared meanings and potential common ground are often missed, and genuine differences are fudged or misrepresented. The language that we use can disguise or obscure the values, assumptions and interests that we bring to the conversation.

And this is not simply a failure of communication. As Bill Murray and Scarlett Johansson discover during their brief encounter in Tokyo, when we skate on the surface of an unfamiliar culture, far more than words can be 'lost in translation'. We fail to spot the tacit clues and signs that allow others to navigate their way around with ease. Uncertain as to the precise source of the problem, we flounder, or look to apportion blame. The Prime Minister, shortly after the general election, pointed his finger at our collective failure to have 'what I call a sensible debate about risk in public policy-making', with the result that 'a mature national conversation on important policy questions like GM science will be impossible'.[30] Sir David King, the government's chief scientist, believes the fault lies with the media. He describes how he has become

> *increasingly concerned at the media's adverse portrayal of science in the news. From the scant media coverage of the MORI poll, which showed the extent to which people support science, to the negative coverage surrounding brain drain . . . these stories continue to run. We have to take action now, and demonstrate that we have a good story to tell, and raise the reputation of science in the media.*[31]

Blind alleys

These are important issues. Few would deny that the media's ways of dealing with science are a problem. But this tells us only part of the story. Both Blair and King are chipping away at a bigger problem: our lack of a shared framework for describing, debating and organising the contribution of science and technology to wider social goals. Without such a framework, we can find ourselves heading up one of two blind alleys.

The first is *determinism*. The political insistence that we must be pro-science and pro-innovation squeezes out any discussion of what sort of science and innovation we want or need. We get locked into a circular discourse in which the only answer to the question 'what is scientific and technological progress?' is 'whatever our innovation systems are delivering'.[32] As we argued in *See-through Science*, just because a new development becomes *possible*, it is usually then seen as desirable. Means have a habit of becoming ends. And before a proper conversation can get under way about the priorities within a particular scientific field, the acceptability of an imagined technology, or the purposes to which it should be directed, policy-makers have already skipped on to the next layer of questions about how to deal with more narrowly defined risks and benefits. Policy and regulatory debates tend to assume that a discussion about ends has already occurred – that the economic and social benefits of innovation are obvious and agreed. But this is rarely the case.

A moment's reflection tells us that no one can be pro-innovation in every sense (do we really want better biological weapons? or human reproductive cloning?), but we lack a framework for dealing with the nuanced and complex set of scientific and technological choices that confront us. Particular trajectories are promoted as if there were no alternatives. All too easily, we fall back into a set of polarised debates in which participants are cast as either 'pro-innovation' or 'anti-science'. There is an assumption that choices which are inherently social and political can be determined by 'sound science'. Yet as Andy Stirling reminds us,

> In reality, science seldom yields such unambiguous answers. Technology in any given field rarely unfolds in only one direction. From the energy sector, through chemicals to food and agriculture, it has been shown time and again that science actually delivers radically divergent answers under different reasonable priorities, questions or assumptions.[33]

The second blind alley is *reductionism*. Even if it is accepted that science cannot be the sole, unproblematic source of authority in these debates, economics is then called on to perform an identical task.[34] Questions about ends and purposes are again airbrushed out, this time to be replaced with the simple calculus of economic growth. We see this tendency in the government's ten-year framework, which constructs the case for more science spending on the twin planks of 'improving the country's future wealth creation prospects' and translating knowledge 'more effectively into business and public service innovation'.[35]

Measuring the contribution of science and technology in primarily economic terms does not rule out the need for forms of public dialogue. Indeed, this can be beneficial as a way of ensuring that 'society's understanding and acceptance of scientific advances moves forward, and does not become a brake on social and economic development'.[36] But, even where dialogue is permitted, another form of reductionism kicks in, as public concerns are invariably framed in terms of risk. The only question we are allowed to ask is 'Is it safe?', with the implication that the likelihood of certain outcomes is susceptible to rational calculation.[37]

Confronted with scientific and technological choices, we need the freedom and opportunity to ask a broader set of questions than economics or risk assessment will allow. And this is where the notion of public value can prove useful: 'if we assume that science's benefits and costs affect citizens in very different ways . . . then public value questions emerge as at least as important as economic ones.'[38]

What is public value?

In 1995, Mark Moore, a Harvard political scientist, published a relatively obscure book on public administration. It put forward the concept of 'public value' as a way of measuring the total benefits – both economic and non-economic – that flow from public policy and investment. Moore and his colleagues were unhappy with the way that traditional theories of public administration treated public managers as robots, who neutrally lent their expertise to whatever purposes were handed to them by politicians or the courts.[39] Instead, he argues that civil servants should 'start to challenge the ends of politics, not just the means', and become 'explorers' who are commissioned by society to use their initiative and imagination in the search for better ways of doing things.[40]

Moore opens his book with the story of a librarian, whose library is being overrun with latchkey children at the end of the school day. The librarian considers introducing new rules limiting children's access, but instead opts for a more entrepreneurial solution. By reorganising the library's layout, and the way she and her colleagues work, she is able to offer a range of improved services, including a children's room, after-school clubs and concerts. As a result, the library is used more often, the children read more books, and the entire community benefits from better facilities. The librarian has succeeded in building public value.[41]

Moore's ideas travelled across the Atlantic, and were eventually picked up by Geoff Mulgan, then head of the Prime Minister's Strategy Unit, who commissioned a report on the role that public value might play in public service reform.[42] The Strategy Unit's report, published in 2002, defines public value as 'the value created by government through services, laws, regulation and other actions'. It provides 'a new language with which to talk about public sector reform with more clarity and certainty', and represents 'a rough yardstick against which to gauge the performances of policies and public institutions [and] make decisions about allocating resources'.[43] And it 'brings concerns about political disengagement into the heart of policy-making, by connecting success measures with the expressed preferences of users'. Value and values are closely linked: 'In a democracy this value is ultimately defined by the public themselves. . . . Public preferences are at the heart of public value.'[44]

In the past three years, the concept of public value has gradually gained currency in debates over the media, the arts and public service reform. Most notably, the BBC organised its Charter Renewal process under the banner of 'Building Public Value'.[45] Think tanks have explored its application to culture[46] and digital policy.[47] Politicians such as Tessa Jowell and Douglas Alexander have advocated it as a way of engaging citizens in public services. One commentator observes, 'In so far as any theory of public management can be, public value is all the rage.'[48]

The appliance to science

So much for the theory. If we import the concept of public value into science policy, can it help us in practice to design better systems of governance? Can it ease the task of translation between scientists, policy-makers and the wider public? Again, it is important to emphasise that this is not simply a matter of new language or better communication. Far more is at stake than language: issues of power, authority, justice and accountability. Nor is it about trying to create a false consensus: no concept is going to provide a point around which all sides can rally. Disagreement and protest, as well as participation, are signs of a healthy democracy. But recognising this complexity, and the conflicts it harbours, should not deter us from the project of developing shared approaches.

Public value provides a route into deliberation about science that avoids the twin pitfalls of determinism and reductionism. Science has major social benefits and thus 'public value'. Yet crucially, as recent controversies have underlined, this value cannot be assumed and taken as automatic, no matter what scientific research is done, or under what conditions. We need therefore to shift from noun to adjective, by asking not only: what is the public value of science? But also, what would *public value science* look like?

Public value also clarifies and deepens the rationale for 'upstream' public engagement. Viewed through a public value lens, engagement might no longer be seen as a 'brake on progress', but instead as a way of maintaining and renewing the social contract that supports science. Upstream engagement enables society to discuss and clarify the public value of science. It encourages dialogue between scientists and the public to move beyond competing propositions, to a richer discussion of visions and ends. And it reminds scientists of the contribution that public *values* can make to the setting of research priorities and trajectories. In the next chapter, we revisit the case for upstream engagement and look at how a range of organisations are approaching this challenge.

3. Upstream – without a paddle?

For 30 years, Ted Freer was an engineering lecturer at Leicester University. But since 1998 he's been engaged in a different sort of science. When Ted's wife was diagnosed with Alzheimer's disease, he started getting involved with the Alzheimer's Society, a charity that puts over a million pounds into dementia research every year.

Alzheimer's disease is relentless. It is a confusing and traumatic experience for patients and the people who care for them. In its final stages, patients can be completely disconnected and isolated, often requiring the full-time care of those closest to them. As with many diseases, most scientific attention is focused on its cause and the possibility of a cure. There is little research designed to help people manage the disease once it has hit.

Ted decided to contribute his experience of caring for his wife to the Alzheimer's Society network Quality Research into Dementia (QRD). Through the network he is one of 150 people directly affected by Alzheimer's who are given the opportunity to influence research into the disease. Contributions from network members are woven into all stages of the research process. They decide priorities, review proposals, interview scientists and monitor research. As a former scientist, Ted realises that the network contributes knowledge and opinions that might otherwise be ignored: 'It provides a totally different viewpoint for researchers . . . if it wasn't for us, they'd only be able to discuss it with their peers.' Initially, the process struck him as unusual, 'But watching it work in practice, I've been reassured. It's very effective.' It also has a therapeutic benefit. Ted's engagement with the research programme helps him to shoulder the burden of caring for someone with Alzheimer's disease. He often talks through research proposals with the people who work at his wife's nursing home.

The QRD network asks important new questions. And a growing number of scientists are rising to the challenge. James Warner is a psychiatrist specialising in dementia at Imperial College in London. In 2002, when he was looking for funding to develop alternative treatments for dementia, he submitted a proposal to the Alzheimer's Society. Having been used to a slow dribble of referees' comments in academia, he was shocked to receive 67 comments from members of the QRD network. But he could see that the comments included some useful pointers that would help to improve his research.

James redrafted the proposal, faced a QRD interview panel and secured his funding. He admits: 'As a one-time cynical scientist, I'm now signed up to the QRD idea. . . . It's not tokenistic. It's real, good quality help.' It has also allowed him to reconnect with his original motivations for doing science. He sees the need for his research to make sense to the people who matter, 'the ones who are shaking the tins in the street'. His initial assumptions about 'us and them' have given way to an appreciation that scientists, patients and carers can work together to improve our understanding of care as well as cause and cure. The QRD network adds an extra layer of expense and complexity to the already tricky process of getting research funding. But these extra costs only upset the balance sheet if we take a very narrow view of the value of science.[49]

Valuing engagement

The QRD network has also given carers a voice in wider discussions of research and treatment. Earlier this year, the National Institute for Clinical Excellence (NICE) recommended withdrawing a clutch of Alzheimer's drugs from NHS treatment. The drugs were seen as over-prescribed and not cost-effective. The Alzheimer's Society, which had earlier helped to convince NICE of the drugs' value, challenged this recommendation. They accused NICE of ignoring the experiences of thousands of carers and patients who benefit from these drugs, in the quest to cut costs. NICE was created to provide an 'evidence-based' assessment of who gets prescribed what. But, as with so much that is justified under the banner of 'evidence', its definition is too narrow to allow it to engage in conversations about what really counts to people. It is tongue-tied when it comes to discussing issues of public value.

Ted Freer knows that Aricept®, one of the drugs in question, is not going to save his wife's life, but it has lessened some of her symptoms and provided a 'window of respite', making it easier for him to care for the woman he loves. These benefits are far removed from NICE's narrow framing of the issue. The voice of the Alzheimer's Society, and its network of carers, is injecting social intelligence into the otherwise desiccated logic of scientific and economic argument. Thanks to the QRD network, at the time of writing, the NICE guidance is being reviewed.

Paddling new currents

The QRD example reminds us that public engagement is not simply about better communication. Institutions need to provide meaningful opportunities for public voices to influence decision-making. In *See-through Science*, we argued that these voices need to be heard early, at a time when they can help to shape scientific priorities. We noted the way that the language of 'upstream' engagement had started to appear in statements by government and the Royal Society. But we warned that upstream engagement would fail if it simply moves the same set of 'downstream', risk-based questions to an earlier point in the research process. Instead, it needs to open up new questions:

> *Why this technology? Why not another? Who needs it? Who is controlling it? Who benefits from it? Can they be trusted? What will it mean for me and my family? Will it improve the environment? What will it mean for people in the developing*

world? The challenge – and opportunity – for upstream public engagement is to force some of these questions back onto the negotiating table, and to do so at a point when they are still able to influence the trajectories of scientific and technological development.[50]

In the past year, the idea of upstream engagement has continued to provoke discussion among scientists and policy-makers. Some have enthusiastically endorsed it, others have diluted it to their taste, and a few have rejected it as unworkable and antiscientific. The government continues to use the term. Speaking about nanotechnology in November 2004, Lord Sainsbury said, 'It is not simply a case of scientists being prepared to engage in debate. . . . If there is one thing we have learnt in recent years it is surely that we need to think about these issues upstream.'[51] Similarly, the journal *Nature*, while acknowledging in an editorial that for some 'the proposal must seem close to giving the lunatics the keys to the asylum', went on to argue:

there are good reasons why scientists should ignore these fears and embrace upstream engagement. . . . Upstream engagement is no panacea. On its own, it won't solve Britain's crisis over trust in science. But it is worth doing – provided that all involved consider two points before beginning. First the process must be long-term and properly funded. . . . More importantly, funding organisations must make a genuine commitment to react to the results of engagement processes.[52]

Stuck in the shallows

Though the concept of upstream engagement has found favour in some parts of the scientific community, the reality doesn't always live up to the rhetoric. It is sometimes portrayed as a way of addressing the *impacts* of technology, be they health, social, environmental or ethical – rather than helping to shape the trajectory of technological development. The hope is that engagement can be used to head off controversy – a prophylactic that we swallow early on and then stop worrying about. There is no recognition that the social intelligence generated by engagement might become outdated or irrelevant as technologies twist their way through the choices and commitments that make up the innovation process.

We see this tendency in the otherwise admirable report on nanotechnologies from the Royal Society and Royal Academy of Engineering. This recognises that 'public attitudes play a crucial role in the realisation of the potential of technological advances',[53] but nowhere suggests that people's values could themselves become the source of alternative research trajectories. The choice we are presented with is advancement or not, faster or slower, but with no real option to change course. This effectively rules out a role for public engagement of a more complex kind, in which scientists and engineers, sensitised through engagement to wider social imaginations, might for themselves decide to approach their science and innovation differently.

Those who see upstream engagement as a means of providing earlier and better predictions of risks and impacts are missing the point. It is not a matter of asking people, with whatever limited information they have at their disposal, to say what they think the effects of ill-defined innovations might be. Rather, it is about moving away from models of prediction and control, which are in any case likely to be flummoxed by the unpredictability of innovation, towards a richer public discussion about the visions, ends and purposes of science. The aim is to broaden the kinds of social influence that shape science and technology, and hold them to account.

Scientism resurgent

As we noted earlier, rumours of the death of the 'deficit model' have been greatly exaggerated. Despite the progress of the science and society agenda, there are still those who maintain that the public are too ignorant to contribute anything useful to scientific decision-making. One of the most vocal is the Liberal Democrat peer, Dick Taverne. In a letter attacking *Nature*'s editorial on upstream engagement, Taverne rejects 'the fashionable demand by a group of sociologists for more democratic science'. He goes on: 'The fact is that science, like art, is not a democratic activity. You do not decide by referendum whether the earth goes round the sun.'[54]

But Taverne is setting up a straw man. As we emphasised in *See-through Science*, upstream engagement is not about members of the public standing over the shoulder of scientists in the laboratory, taking votes or holding referendums on what they should or should not be doing.[55] That Taverne can conceive of accountability only in these terms reflects nothing more than the poverty of his own democratic imagination. This agenda is not about imposing cumbersome bureaucratic structures on science, or forcing lay people onto every research funding committee. Questions about structures do need to be considered, but are a sideshow compared with the far more important – and exciting – challenge of building more reflective capacity into the *practice* of science. As well as bringing the public into new conversations with science, we need to bring out the public *within* the scientist – by enabling scientists to reflect on the social and ethical dimensions of their work.

We need to break down some of the false oppositions between scientists and the public that critics such as Taverne seek to perpetuate. Those scientists who take part as expert witnesses in public engagement exercises, such as citizens' juries, are frequently surprised at the insight and common sense that ordinary members of the public bring to such interactions. At its most effective, upstream engagement can help to challenge the stereotypes that scientists and policy-makers have of the public. But it is important to start by wiping the slate clean of assumptions about who the public are and what they think.

The end of the line

When all else fails, critics of upstream engagement tend to resort to arguments based on a linear model of innovation. They grudgingly concede that technologies and applications may merit some public discussion, but insist that 'basic science' should be kept apart, as a unique domain governed by curiosity and 'science for science's sake'.

Yet like deficit models of the public, linear models of innovation are a default, unthinking response to the complexity of the subjects they purport to describe. As John Ziman observes, despite the fact that 'the linear model of technological innovation is obviously over-simplified . . . it underlies what most politicians, business people, civil servants and journalists say about science'.[56] Rhetorics of linearity have created some powerful myths in the popular imagination,[57] such as the idea that the discovery of new knowledge is the basis of innovation, or that science and technology are inevitable, but distinct, points on the same line. There is no questioning of *what* science, or *which* technology – the only issue is how fast we can move from one end to the other.

But of course, innovation doesn't happen in a line. Successful technologies are the products of networks of interaction between inventors, scientists, engineers, users and business people. For every technology that seems to spring from a clear advance in our scientific understanding, there are others that are prompting new lines of scientific enquiry, or whose powers of observation, calculation and analysis are allowing new types of science to be done.

Basic research is an attractive idea. And the idea of 'science for science's sake' may still motivate many researchers. However, the intellectual curiosity of individual scientists does not aggregate into the structures, expectations and funding opportunities that shape collective patterns of research. Scientists who would consider themselves to be doing basic research are frequently asked by funding bodies to justify their work in terms of its possible future benefits to society. As Helga Nowotny has described: 'It is not Nature whispering into the ears of researchers which problem they should address next, but an intricate mixture of opportunities and incentives, or prior investments and of strategic planning mixed with subversive contingencies.'[58]

If we visited a nanotechnology lab, we might observe an experiment designed to understand how nanoparticles could improve drug delivery. Or an attempt to develop longer carbon nanotubes. In both cases, the researchers would be relying on highly advanced microscopes. In such a setting, what counts as basic science and what as technology? As one top nanoscientist says:

The basic science argument is a red herring. You look at an EPSRC application form and there's a section on outreach, applications and expected patents. There's almost no such thing as basic science anymore. Maybe if you're working in theoretical mathematics or particle physics, the idea has some meaning. But for everyone else, we now find that the science in the lab one week can pretty much be in the shops the week after. The process from basic science to applications has accelerated enormously.[59]

This is not a recent development. Historians of molecular biology have identified the critical role that social visions, about genetic improvement and the engineering of artificial life, played in the framing of scientific research from a very early stage. For example, Lily Kay, documents how the enormous investments in this research by the Rockefeller Foundation, through its 'Science for Man' programme, had clear visions of the social ends to which the knowledge would be directed:

From its very inception around 1930, the molecular biology scientific programme was defined and conceptualised in terms of technological capabilities and social possibilities . . . the ends and means of biological engineering were inscribed into the molecular biology programme. . . .

Scientists and patrons came to share a molecular vision of life. . . . Though not an applied science, molecular biology in the 1930s and 1940s was mission-oriented.[60]

So what would commonly be called basic science, and assumed to be free of any social influences, was in fact pervaded with visions of the future. Such visions may not have impinged directly on scientists' daily work, but they did play a profound role in shaping research cultures and expectations. Most crucially for our argument about upstream engagement, they were imposed on society with no debate.

The cycle of engagement

In criticising the linear model, we also need to acknowledge the linearity of our metaphorical stream. A limitation of the notion of 'upstream' engagement is its implication that we can move up and down innovation processes at will, inserting a bit of public engagement where we judge it will be most effective.[61] Rather than opening up innovation to alternative trajectories and possibilities, such an approach would risk closing it down, by restricting public engagement to a specific point in the process.

The 'upstream' metaphor is further proof of the powerful hold that linear models exert on all our imaginations. But needless to say, we do not mean it to imply a one-off fix: that we can do public engagement early but not often. Rather, upstream engagement – at a point where research trajectories are still open and undetermined – should be the start of a process of *ongoing* deliberation and social assessment, that embeds dialogue between scientists, stakeholders and lay publics within all stages of the R&D process. A recent paper by Roland Jackson and colleagues includes a diagram which neatly conveys this idea of an ongoing cycle of engagement (see figure 1).[62]

Jackson and colleagues explain the diagram as follows:

> *If we imagine a cycle . . . it seems evident that different models of engagement are suitable at different stages. In general, where the research is in early stages and especially where it is leading-edge and complex and there is great scientific uncertainty about outcomes, benefits and risks, small scale deliberation between scientists and others will tend to be most appropriate. Once applications and consequences are more evident, either anticipated or already realised, mass participation methods become more relevant.*[63]

It is also important to acknowledge that the history of innovation is full of cases where it is difficult to envisage when, and in what ways, public engagement might have been useful. For example, Sir John Enderby, Vice President of the Royal Society, has suggested that any attempt to engage the public in an upstream discussion about laser

Figure 1 When and how should public engagement take place?

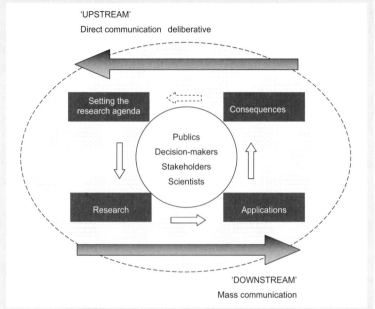

'UPSTREAM'

Direct communication   deliberative

Setting the research agenda

Consequences

Publics
Decision-makers
Stakeholders
Scientists

Research

Applications

'DOWNSTREAM'

Mass communication

Source: Jackson et al, 'Strengths of public dialogue on science-related issues'

technology would have been pointless.[64] For many years, the laser appeared to be a solution without a problem. Charles Townes, the US scientist who won a Nobel prize for his contribution to its discovery, was regularly teased by colleagues about its seeming irrelevance to the real world. The many applications of the technology we use today, from DVD players to eye surgery, were only realised many years later.

Had they been invited to give an opinion, what might ordinary members of the public have said about the laser in the 1950s? If this example tells us anything, it is that innovation is not inevitable or straightforward. Encouraging early debate is never easy, but this is no reason not to try. Who knows? If there had been more public engagement about the laser in the 1950s, some of its more valuable applications might have reached us a little sooner.

Putting it into practice

A growing number of organisations are experimenting with new forms of public engagement. We have already described the approach the Alzheimer's Society is taking, but here we offer three more snapshots of efforts to 'paddle upstream'. These are all embryonic processes, and their eventual outcomes and effects cannot be predicted. But they are important sites of social learning, from which further improvements and productive questions can be developed. In each of these organisations, theories about public engagement and public value science are being tested and refined. If these debates are to move forward, we can't afford false pretences – including the false pretence that this will all be easy.

The MRC's Advisory Group on Public Involvement

Medicine has always been conversational. To be effective, it requires engagement, first and foremost with patients, but also with wider networks of stakeholders and the public. Among the UK research councils, the Medical Research Council (MRC) has long been at the forefront of efforts to involve the public. In 2000, it created a Consumer Liaison Group. This changed its name in 2004 to the Advisory Group on Public Involvement (AGPI), to reflect the more strategic and proactive role that the group was starting to play in MRC decision-making.

AGPI members are drawn from all walks of life. Elizabeth Mitchell, the MRC's external communications manager, describes them as bringing a 'public voice' into decision-making: 'They are not representative of the public in any formal sense. But they bring a breadth of perspectives that help us to reflect wider public views.'

How does this public voice get heard in the decisions that matter? Mitchell admits that it isn't always easy to connect AGPI with other structures within MRC. But she is optimistic that it is having an effect: 'AGPI has changed the culture within MRC. We think a lot more now about when and how to involve the public.' In the past year, AGPI members have started to attend the MRC's main research boards, which decide strategy and priorities: 'We are trying to operate upstream,' says Mitchell. 'We want to get more involvement at an early stage in key discussions.'[65]

From the autumn of 2005, Colin Blakemore, MRC's chief executive, will take over as Chair of AGPI's meetings. This is a sign of the growing importance that the MRC is placing on public involvement, and reflects a commitment to ongoing experimentation. 'I'm very committed to engaging the public,' says Blakemore. 'However, in the rush to openness and transparency, we need to think carefully about what models will work best for scientists, and what works best for the public.'[66]

Civil society meets Framework Programme 7

Each year, the European Union makes significant investments in R&D, through its Framework programmes. Framework 6, which is currently running, has a budget of €17.5 billion, and this is proposed to increase to nearly €40 billion under Framework 7. The choices and priorities that shape these investments have not usually been the subject of stakeholder or public discussion.

But as Framework 7 was being developed, a group of NGOs and civil society organisations launched a campaign for greater account-ability and transparency. They began circulating a petition, which argued that 'the current proposals for developing FP7 place too much power in the hands of the industry lobby and not enough influence from the wider European public in whose name this money is being spent'.[67] Instead, they called for the themes of Framework 7 to be recast towards social, environmental and public health goals.

Claudia Neubauer, who works for the French NGO Fondation Sciences Citoyennes, was one of the group behind the petition. She explains:

We decided to build an informal network, but what we discovered was that most of the big NGOs don't campaign on research policy, even though they may spend a lot of their time addressing issues that are the result of research decisions made 20 years ago. . . . Just as governments and debates about risk are moving upstream, so civil society is now moving upstream.

Will such moves help to avoid conflict at a later stage? Neubauer sees an important role for both protest and participation: 'Direct clashes can help to mobilise people, but the idea of upstream involvement is to try and also put a positive agenda onto the table at an earlier stage.'

Engagement can take many forms. This is an example of people on the outside calling for particular issues to be placed on an organisation's agenda. The organisation in question may respond by closing its ears or turning away. But in this case, there are some positive signs that the European Commission is listening. Neubauer explains: 'Through this campaign, we came into contact with the people inside the Commission working on these issues, and they have been largely supportive. It is not a homogeneous institution.'

The Commission has now set up a working group on 'Science and Governance', and invited Neubauer to participate as a civil society representative. She hopes this group can become a catalyst for greater change. But she is also realistic about the obstacles: 'Of course, we don't know yet if any of our proposals will be incorporated into the next round of Framework 7 proposals. This will be the real test. We shall see.'[68]

Public engagement on nanotechnologies in the UK

One of the principal recommendations of the Royal Society and Royal Academy of Engineering report on nanotechnologies was for a 'constructive and proactive debate about the future of nano-technologies [to] be undertaken now – at a stage when it can inform key decisions'.[69] A year later, the conversation is getting under way. The OST has recently published an overview of the engagement activities that it is supporting.[70] These include the NanoDialogues project, which involves Demos and Lancaster University (see box), but also two other projects:

- ○ Small Talk – a joint initiative by the BA, Royal Institution, Think Lab and others, which involves a series of events and discussions on nanotechnologies across the UK
- ○ Nanotechnologies Engagement Group – a network, coordinated by Involve, which is designed to support public engagement activities, and foster links and learning from one project to another.

The NanoDialogues: experiments in public engagement

Over the next 12 months, with support from the government's Sciencewise programme, Demos and Lancaster University will be facilitating a series of practical experiments in public engagement, designed to inform decision-making around nanotechnologies.

Experiment 1 – Nanoparticles and upstream regulation

Partner: Environment Agency

Working with the Environment Agency, our first experiment asks how discussions between regulators and the public can contribute to sustainable innovation and regulation of nanotechnologies. The Environment Agency is a firm advocate of 'risk-based regulation', but the uncertainties surrounding nanoparticles in complex ecosystems make risk assessment very difficult. Through a citizens' jury, to be held later this year, the Agency will invite public input to emerging thinking about nanoparticles, regulation and environmental remediation.

Experiment 2 – Imagining publicly engaged science

Partners: Biotechnology and Biological Sciences Research Council (BBSRC) and Engineering and Physical Sciences Research Council (EPSRC)

Research councils are a key influence on what is considered to be valuable science. The aim of our second experiment is to investigate the potential for public debate at an early stage in decision-making. Working with two of the research councils – BBSRC and EPSRC – we will be exploring what might be at stake as biotechnologies and nanotechnologies converge. How can dialogue between scientists and the public clarify key questions? And how can research priorities reflect public concerns?

Experiment 3 – Nanotechnologies in development

Partner: Practical Action

Too often, the voices of people in developing countries are neither sought nor taken account of in decisions about science and innovation. For our third experiment, we are working with Practical Action, the development NGO, which has a lot of experience of public participation in developing countries. Practical Action will facilitate discussions with two community groups in South Africa about the potential contribution of nanotechnologies to the provision of clean drinking water. Several UK nanoscientists will be invited to participate, as a way of deepening their understanding of local contexts, priorities and needs, and the implications these might have for their research.

4. Citizen scientists

An ambitious young scientist may dream of one day running his own lab. He is unlikely to imagine that he will end up working with Greenpeace. But Mark Welland, who runs Cambridge University's nanoscience centre, has recently been engaged in a novel experiment. Together with Greenpeace, the *Guardian* newspaper and Newcastle University, he has initiated the UK's first citizens' jury on nanotechnology. Over several weeks, the jury of 20 men and women from Halifax in West Yorkshire heard evidence and reflected on the potential implications of nanotechnologies.

Mark admits that, for him, it has been a pretty steep learning curve: 'Ten years ago . . . terms such as citizens' jury, public engagement and democratisation of science were grouped together in my mind as a science fringe activity largely patrolled by pressure groups.'[71] As discussions of the social implications of nanotechnology began to gather momentum, he became more directly involved. He joined the working group of the Royal Society and Royal Academy of Engineering inquiry, and hired a social scientist to work in his lab, to help him and his team to understand these concerns.

He is not yet convinced that exercises such as citizens' juries are the best way of doing things, but his willingness to learn shines through: 'I'm not sure yet what's going to be effective in the long term, what will really work. There are some people in the science community who aren't happy that I'm doing this, who say I should get on with the science.' But he remains confident that it is worthwhile: 'For us, the jury is a small step in the right direction. We've gone into it openly and honestly. There's nothing to hide from. It's a very positive way of supporting our science.'[72]

At the other end of the nanoscience career ladder sits Alexis Vlandas, a 24-year-old postgraduate at Oxford University. What really turns Alexis on is the science: 'Most of the scientists I know are motivated by pure science, by discovery. I'm thrilled by the questions I'm addressing – how the properties of materials change when we shrink them.' While he spends his days in the lab seeking answers to these questions, he also grapples with some bigger issues outside. He coordinates the Oxford branch of Pugwash, the international network for socially responsible scientists, and regularly organises meetings on the social dimensions of nanotechnology or the dilemmas raised by corporate and military funding of research. He has also written a paper for a social science journal, on comparisons between public debates around biotechnology and nanotechnology.

In fact, Alexis does a lot that falls outside the prevailing model of what counts as good science:

> One of the senior scientists in my group said to me 'what kind of scientist do you want to be, a social scientist or a real scientist?' He's worried that I spend too much time on this stuff. . . . If you want to get a place at Harvard or any other top university, there's no incentive to engage in these debates . . . the funding system infantilises you as a scientist. It pushes scientists to steer their research to fashionable areas. There's no space to say 'Wait a minute, should we be asking different questions?'

When Alexis gets his Oxford DPhil he will become a paid-up member of the scientific establishment. This won't stop him from continuing to reflect and act on these issues: 'As a responsible scientist, I think you have to engage in the wider debate. . . . If you think your research will ultimately be used to improve missile technology, this is a big issue for me and many scientists.' However, Alexis has had to do this in his spare time, as a hobby. He has been disappointed by the lack of attention paid to such issues within his course:

> It seems strange to me that at one of the UK's top universities, you have compulsory courses on attracting venture capital and business angels, but nothing on the history of science, the philosophy of science, the social impacts and dilemmas of technology. I'm genuinely surprised by this. What does it say about our value system? What signals does it send to younger scientists?[73]

Some critics have argued that public engagement leads to the 'demoralisation of scientists'.[74] But Mark and Alexis are far from demoralised. Rather, they seem invigorated by reflecting on the social dimensions of their work. Ten years ago, the sociologist Alan Irwin published *Citizen Science*, which explored the relationship between science, the public and the environment. Irwin asked important questions about our notions of science and of citizenship, and argued that we need to rethink the relationship between science and everyday life.[75] For the next phase of the science and society agenda, the most pressing challenge is how to strengthen the contribution that scientists and engineers can themselves make to the health and robustness of our shared public realm.

Both Mark and Alexis represent what we might call *citizen scientists*, who increasingly treat reflection on these issues as part of their everyday work and responsibility. Yet the structures that surround them – for funding, research assessment and career development – often push in the opposite direction. The question is how we can broaden our notions of scientific excellence to support such activities and make them part of the normal practice of doing good science.

Cultures of science

In her presidential address at last year's BA Festival, Julia Higgins reminded us that we often talk about 'scientists' with one image in mind: the lab-based researcher. In fact, there are many types of scientist and engineer who can potentially contribute to the building of new cultures of responsibility:

- O people whose profession requires scientific training, such as doctors and science teachers
- O people who are appointed to advisory positions as scientists, such as the chief scientific adviser and the heads of the research councils
- O people in influential positions who are known to be scientists, for example in business or Parliament
- O people who have scientific training but are not explicitly using it in their professions, such as accountants, civil servants and managers
- O and finally, the group we think of most often: research workers in academia and industry.[76]

Higgins went on to ask what makes scientists tick: What regulates them and what motivates them? What keeps them in their place or allows them to progress? The most influential account of the values that keep science productive, organised and respected came from sociologist Robert Merton in 1942.[77] Merton suggested that the 'prescriptions, proscriptions, preferences and permissions' that scientists feel obliged to follow could be summed up in a small number of norms. These are commonly known by the acronym 'CUDOS'. Good science, and the kudos that comes with it, is seen to rely on these five norms. *Communalism* refers to the responsibility to share all knowledge claims, data and experimental design within a community of peers. This is evaluated according to standards which are *Universal* (uninfluenced by local context or bias) and *Disinterested* (removed from any personal interests except a commitment to valid knowledge). Good science is also *Original* and its validity is tested through organised *Scepticism*, whereby no claim is accepted unless it passes independent tests of replication.

Several decades of empirical research by sociologists of science have shown that Merton's norms fall short as a descriptive account of the work that scientists do. Especially in the period during the war (when Merton published) and just after, communalism was accompanied by a hefty dose of secrecy. And in many areas of science, especially where companies and the military are involved, this secrecy remains today. The organised scepticism that Merton longed for, enshrined most visibly in peer review, has also been found to be tainted by commitments to certain people or patterns of thought.[78] And in areas of strong disagreement between opposing schools, studies have shown that there is unlikely to be any agreement on what counts as an adequate or fair replication of an experiment.[79]

So in practice, even academic science appears to transgress these norms, and this is before we consider the other influences, such as the relentless need to publish papers, to secure research funding or, particularly for contract researchers, to create some semblance of job security. The norms' lack of descriptive power does not prevent them from serving a useful purpose. But we need to face up to some serious tensions within them. For example, the competition engendered by scepticism and a need for originality can stimulate important, dynamic science, but can also lead to greater secrecy. Similarly, science that tries too hard to be disinterested might prevent uncertain new areas from being explored, and might extinguish some of the passion that scientific research should inspire.

Merton's norms assumed a certain type of public value. Good science was seen as good for society – 'speaking truth to power' was its leitmotif. But how useful are the norms now, in an era of citizen science? Should we, for example, be trying to foster diversity in science as well as disinterestedness? How might we develop ways in which this diversity can be collaborative and interdisciplinary rather than competitive? How can science remain disinterested without alienating its publics? And how can we recover some of the passion and excitement of excellent science? As science comes to terms with its place in modern societies, it needs to adapt its norms to incorporate new approaches and models. We would suggest that these should include concepts such as upstream public engagement, and what Dave Guston and Dan Sarewitz have termed 'real-time technology assessment'.[80]

Julia Higgins agrees that we should be prepared to rethink and broaden our definition of good science: 'Do we mean exciting and novel? Careful and thorough? Safe? Probably yes to all these. But what about relevant? Applicable? . . . I think it is a pity we do not spend more time openly discussing what we mean by "good science".' To develop new frameworks for excellence we need to interrogate public values rather than assume them. We need to acknowledge and encourage the many varieties of social intelligence that can contribute to science. Some of this will come from other scientists and engineers, who can contribute to debates outside their own area of research; some from policy-makers, NGOs and other stakeholders; and some from ordinary citizens.

In Mark Moore's theory of public value, the focus is on those responsible for managing public services. They are the people who have to identify and build new forms of public value. He emphasises that they must have the right incentives and freedoms if they are to do this effectively. The same applies in science. If scientists and engineers are to build public value, we need to take seriously the constraints and choices that they face, especially during the early stage of their careers. This is an issue that the Royal Society is examining over the next year, through a study of scientists in more than 50 universities, which will seek to identify the factors that facilitate or inhibit science communication and public engagement.[81]

In recognition of the diversity of people involved in scientific work, we need to think hard about which groups are best placed to create public value, and in what ways. But if social reflection and public engagement are not rewarded, or even acknowledged, by the cultures and incentive structures that these different scientists inhabit, we will be forced to rely on the efforts of determined individuals such as Mark Welland and Alexis Vlandas to bridge the divide between science and the rest of society.

5. Challenges great and small

Do we need a fresh set of ambitious goals for science? Where are the inheritors to the sense of purpose that drove the Apollo space programme, Nixon's 'War on Cancer' or the Manhattan Project? Deep in the bowels of Whitehall, a group of senior civil servants are grappling with precisely this question. The Coordination of Research and Analysis Group, or CRAG, which consists of the government's chief scientific adviser, chief economist and chief statistician, plus a handful of top strategists from Number 10 and the Cabinet Office, is developing a set of 'grand challenges' – primarily scientific and technological, but also economic and social – that can help to orient policy and investment decisions for the next decade.

It is an idea that others have used to powerful effect. In 2003, the Bill and Melinda Gates Foundation launched a call for proposals under the banner 'Grand Challenges in Global Health'. This generated over 1000 proposals from scientists and health experts around the world, and in June 2005, 43 of these were awarded a total of US$450 million in funding. Among the ideas that have received support are vaccines that do not require needles or refrigeration, and new ways to prevent the transmission of malaria: 'We were amazed by the response,' said Harold Varmus, who chaired the selection board. 'Clearly there's tremendous untapped potential among the world's scientists to address diseases of the developing world.'[82]

Back in Whitehall, it is not yet clear if this is the kind of approach that CRAG has in mind. One official close to the process admits that the grand challenges are 'really just a marketing term' for cross-cutting initiatives that have a science or technology dimension. But the aim is to crystallise the challenges in time to influence the negotiations set to begin in 2006 on the next spending round, which gets under way in 2008. Accelerating the UK's transition to a low-carbon economy is likely to be one major focus.

Governments rarely find it easy to plan for the future. As Geoff Mulgan has observed, they suffer from an 'optical distortion' which leads them to overestimate what can be changed short term, and underestimate just how much can be changed over the longer term. This pattern was reflected in Labour's 2005 election manifesto, which was 'full of detail, careful preparation and ambition aimed at the three-to-five year time horizon. . . . But it did not attempt a longer perspective and excluded several of the most challenging and obvious issues, such as pensions reform and energy policy.'[83]

So for the government to have embarked on identifying a set of long-term challenges is a positive and encouraging sign. In this same spirit, we would lay down a grand challenge of our own. Can this process mark the start of a wider debate about the public value of science? Will it discuss the role of scientific innovation in addressing society's problems, alongside other forms of cultural, political and institutional innovation? And can such discussions be opened up to include a broader range of voices and perspectives? If this process is to be used to identify long-term research priorities, it should also set a new benchmark for stakeholder and public engagement.

At the same time, let no one pretend that grand schemes – however well conceived or widely debated – can resolve the deeper, more systemic challenges that confront science: issues of governance, culture and accountability that have moved up the scientific agenda in recent years, but now require a more determined response. In this final chapter, we highlight two areas where we believe efforts should be targeted – *research cultures* and *policy capacity* – and we address a number of specific policy questions.

Research cultures

The Research Assessment Exercise

Everyone accepts that there needs to be some way of evaluating and comparing research performance, but the Research Assessment Exercise (RAE) has few friends. In July 2005, the Higher Education Funding Council published the guidelines for the 2008 RAE, and many researchers will have spent the summer picking over the small print. Yet the intense competition that the RAE unleashes, its distorting effect on research priorities, and the sheer bureaucracy of the undertaking fills many researchers with despair. In 2008, 900 academics on 67 subject panels will evaluate the outputs of their peers. Each university department will be graded according to its outputs, environment and esteem – results which then determine future funding. Already, the jostling for position in 2008 is well under way, with universities head-hunting star names to boost their key departments.

How does the RAE relate to the science and society agenda? If we are serious about creating more open and reflective research cultures, we need to recognise the barriers. The RAE creates no incentive for university scientists to devote time and energy to these issues. Rather, it reinforces the model of the highly specialised researcher, locked in a cycle of publish-or-perish. The 2008 guidelines do recognise a few other activities: for example, historians can now get credit for involvement in a TV programme. But wider efforts to contribute to policy or engage in dialogue with the public are still unrewarded. This undermines efforts by the research councils to encourage more engagement and social reflection. The criteria that the RAE is based on are far too narrow. In particular, assumptions of what constitutes good research and good science need to be revisited and redefined.

Research councils

Turning to the other half of the dual support system – the research councils – here the picture is much brighter. Each of the councils appears to be taking public engagement seriously, and although they move forward at different speeds, there are many initiatives and experiments under way. A number of the Councils (notably BBSRC, EPSRC and MRC) are looking explicitly at what upstream engage-ment might mean in the context of their work.

One area that would benefit from more thought is the approach that the research councils are taking to interdisciplinary work in this area. The nanotechnologies report from the Royal Society and the Royal Academy of Engineering emphasised the need for novel forms of interdisciplinary collaboration to explore emerging social and ethical issues.[84] The experience of Mark Welland's Nanoscience Centre in Cambridge, with its resident social scientist, demon-strates the practical value such collaborations can bring, by supporting ongoing reflection by research scientists on the social dimensions of their work.[85] Yet for natural and physical scientists, social scientists and engineers seeking to embark on similar initiatives, there are no obvious funding mechanisms. Such activities tend to fall outside the remit of individual councils, and there needs to be more support for research that works across these boundaries.

Public–private research

A final point concerns the growing influence of the private sector on university research. In a penetrating new study of the US higher education system, the journalist Jennifer Washburn charts the effects of there being ever closer ties between the public and private research sectors. Echoing the sentiments of Steven Rose, whom we quoted earlier, Washburn concludes:

> *Market forces are dictating what is happening in the world of higher education as never before. . . . Universities now routinely operate complex patenting and licensing operations to market their faculty's inventions. . . . The question of who owns academic research has grown increasingly contentious, as the openness and shaping that once characterised university life has given way to a new proprietary culture.*[86]

The US is almost certainly a few years ahead of the UK in terms of these trends, but the thrust of the government's ten-year framework and the Lambert Review is to accelerate and multiply public–private collaborations wherever possible.

We would emphasise that we are not opposed to this in principle. Collaboration between universities and businesses can be very positive, and there are strong economic arguments why the UK needs a lot more of it. Also, there never was a halcyon day when public science took place completely unsullied by private sector influences. Even an iconic scientific figure such as Galileo routinely integrated monetary and utilitarian interests with his 'natural philosophy', as historians of science have shown.[87]

The question is not *if* we strengthen such links but *how*. Can we do it in a way that maintains the openness and integrity of academic research cultures? In what ways will an increasing role for business in university life support or impede efforts to move research cultures in a more socially reflective and publicly engaged direction? Under what conditions can private sector investment generate public value, and when might it undermine it?

Reviewing Washburn's book in the *Financial Times*, Alan Ryan, the warden of New College, Oxford, was compelled to wonder 'what a British version of Washburn might uncover. British universities have lately been encouraged to engage in aggressive patenting and licensing and it is hard to believe that they do not run the dangers she describes.'[88] There is a pressing need to examine some of these tensions and discuss them honestly, rather than pretend that no such problems will ever arise. This area would benefit from more detailed analysis and scrutiny by the House of Commons Science and Technology Committee. Such an inquiry could also incorporate some of the questions about military influences on university science and technology that were raised in a recent report by Scientists for Global Responsibility.[89]

Policy capacity

A commission for emerging technologies and society

Rapid advances within – and at the intersections between – nano-technologies, biotechnologies, information technologies and neuro-science are giving rise to new, and potentially profound, economic, social and ethical questions. Confronted with such rapid tech-nological change, how should individuals and institutions imagine new social possibilities and choose among them wisely? A recent report from the European Commission highlights some of the difficulties that we face:

> *The convergence of these profoundly transformative technologies and technology-enabling sciences is the major research initiative of the twenty-first century. If these various technologies created controversies and anxiety each on their own, their convergence poses a major challenge not only to the research community, but from the very beginning also to policy-makers and European societies.*[90]

Policy responses tend to be defined primarily in terms of narrow technological categories (witness, for example, the number of new committees that have been set up by government to deal with nanotechnologies in the past year). But by framing the problem in this way, we can lose sight of the more fundamental questions that arise in relation to almost all new technologies. It is easy to miss opportunities for social and policy learning from one technological 'episode' to the next. For example, debates over nuclear power in the 1960s and 1970s profoundly shaped responses to GM crops in the 1990s, and the GM controversy in turn shaped the recent reception of nanotechnologies. But where are the opportunities for systematic reflection and policy learning across these different domains?

One of the most thoughtful and widely praised recommendations of the Royal Society and Royal Academy of Engineering report on nanotechnologies came at the very end of the Commission's report.

Recommendation 21 called on the government's chief scientific adviser to 'establish a group that brings together representatives of a wide range of stakeholders to look at new and emerging technologies'.[91] This was interpreted by many at the time to imply a new committee of some sort, made up of a mix of business, policy, stakeholder and public representatives, that would provide government with independent advice.

Yet the government's response has been to allocate this task to a new horizon scanning centre that was already being established within OST (having been previously announced in the ten-year framework).[92] We are confident that this horizon scanning centre will perform some useful and valuable work. But we are concerned that this nonetheless represents a significant watering-down of what was envisaged in the Royal Society and Royal Academy of Engineering report.

Earlier this year, the Agriculture and Environment Biotechnology Commission (AEBC) was wound up. During its five-year lifespan, the AEBC worked on a range of issues and provided a new model for inclusive, independent scientific advice. The diversity of its members, drawn from all sides in the debate, meant that it was widely respected. With its demise, we have lost an important voice in British science policy. We believe that government should build a more radical and wide-ranging body from the ashes of the AEBC, that can fulfil the spirit of Recommendation 21 and advise on the long-term implications of new and emerging technologies.

A Commission on Emerging Technologies and Society would make an important contribution to the aspirations set out in the ten-year framework 'for the UK public to be confident about the governance, regulation and use of science and technology'.[93] It would improve our social foresight, and also our hindsight, by encouraging systematic learning from recent experiences with technology. It would provide an institutional home for 'real-time technology assessment', of the type that is now getting under way in the US for nanotechnologies.[94] Above all, it would reflect the need for new spaces where genuine dialogue and learning can take place between policy-makers, scientists, social scientists, NGOs and wider publics. Half of the Commission's members could be drawn from the worlds of science, policy and other stakeholder groups, and half from the general public, to make it a melting pot for different views and perspectives. It could also initiate and oversee wider public engagement exercises – becoming both a hub and a broker for activities in this area.

Progressive globalisation

Few ministerial speeches about science and innovation are now complete without an obligatory reference to China and India. These two vast, heterogeneous nations – home to a third of the world's population – are perpetually conjoined in a form of political shorthand designed to convey the onward march of globalisation.

We also need to be alert to the way these new 'science powers' are used to argue for a more relaxed stance on social, ethical or environmental issues here in the UK. Tony Blair's speech to the Royal Society in 2002 is a notable example:

> *The idea of making this speech has been in my mind for some time. The final prompt for it came, curiously enough, when I was in Bangalore in January. I met a group of academics, who were also in business in the biotech field. They said to me bluntly: 'Europe has gone soft on science; we are going to leapfrog you and you will miss out.' They regarded the debate on GM here and elsewhere in Europe as utterly astonishing. They saw us as completely overrun by protestors and pressure groups who used emotion to drive out reason. And they didn't think we had the political will to stand up for proper science.*[95]

Those of us who advocate more socially responsive and accountable forms of science and innovation need to take this 'Wild East' argument seriously. But we believe it is possible to mount a robust response. Our first defence has to be that this is a counsel of despair, the logical end point of which is a set of lowest-common-denominator standards not just for science, but also for labour rights, civil liberties and environmental standards. Just as on these other issues, there is a clear progressive case for public value science. It is also misleading, not to mention deeply patronising, to pretend that people in India and China don't share many of these same concerns – albeit expressed in a variety of ways.

In his latest book, *The Argumentative Indian*, Amartya Sen offers a colourful account of the role that public reasoning, dialogue and debate have played through India's history. He effectively dispels some of the myths and stereotypes of India as a land of exoticism and mysticism, or the new high-tech, back office of the global economy. And he reminds us that although it has not always taken a Western representative form, there is this deep seam of democracy, or 'government by discussion' running through Indian culture. He tells the story of how, just before the Indian general elections in the spring of 2004, he visited a Bengali village not far from his home, and was told by an elderly man who was barely literate and certainly very poor: 'It is not very hard to silence us, but this is not because we cannot speak.'[96]

So, although it is often claimed that democracy is a quintessentially Western idea and practice – with a direct lineage running from ancient Athens to the White House – such a view neglects the many varieties of public discussion and public reasoning that have always existed in India, and exist today in most cultures. Even in China, where there is less freedom to debate such issues in formal terms, the environmental and social consequences of rapid technological development are now becoming the focus of intense political debate, and at times public protest.[97]

The way our politics describes the relationships between science, globalisation and competitiveness must start to reflect these subtleties. Instead of seeing the UK's progress towards more democratic models of science as a *barrier* to our success in the global knowledge economy, can it not become a different form of advantage? Might it not lead us down new – and potentially preferable – paths of innovation? The evidence we have from the environmental sphere suggests that countries can gain competitive advantage from the adoption of higher standards.[98] We need to explore whether similar patterns can emerge here. There may also be insights from scientific governance, ethics and public deliberation that we can exchange and export. We need to develop networks that allow policy-makers and scientists in Europe to forge common purpose and alliances on these issues with their counterparts in Asia.

These are difficult issues and we do not pretend they can be easily resolved.[99] But they bring us back to where we started: the fundamental questions of why we do science, where it is taking us, and who it is for. Tony Blair's speech to the Royal Society, in which he warned of emotion driving out reason, was titled 'Science matters'. Our argument has been that, yes, science does matter. But it matters for more than narrow, economic reasons. We need to talk, and occasionally to argue, about why this is so. And we need to infuse the cultures and practices of science with this richer and more open set of social possibilities. This is how, together, we can build public value.

Notes

1   'Save British Science', advertisement in *The Times*, 13 Jan 1986.
2   Interview with Peter Cotgreave.
3   HM Treasury/Department for Trade and Industry/Department for Education and Skills, *Science and Innovation Investment Framework 2004–2014* (London: HM Treasury, Jul 2004).
4   T Blair, 'Science matters', speech delivered at the Royal Society, 23 May 2002; see: www.number10.gov.uk/output/Page1715.asp (accessed 13 Aug 2005).
5   Interview with Peter Cotgreave, 2 Aug 2005.
6   House of Lords Select Committee on Science and Technology, *Science and Society* (London: House of Lords, 23 Feb 2000).
7   The Royal Society and Royal Academy of Engineering, *Nanoscience and Nanotechnologies: Opportunities and uncertainties* (London: The Royal Society/RAEng, Jul 2004); HM Government, *Response to the Royal Society and Royal Academy of Engineering Report* (London: HM Government, Feb 2005).
8   MORI/DTI, *Science in Society – Findings from qualitative and quantitative research* (London: DTI, Mar 2005).
9   Anonymous interview, Jul 2005.
10   House of Lords Select Committee on Science and Technology, *Science and Society*, see chapter 1, paragraph 19.
11   For example, Dick Taverne and colleagues in the group 'Sense About Science'.
12   D Derbyshire and R Highfield, 'US attacks "contributed to mistrust of science"', *Daily Telegraph*, 9 Sep 2002.
13   T Radford, 'Britain is warned of "scientific dark ages"', *Guardian*, 8 Sep 2003.
14   J Higgins, 'Responsibility and science', presidential address to the BA Festival of Science, 6 Sep 2004.
15   A Broers, 'Risk and responsibility' (lecture 5), *The Triumph of Technology*, Reith Lectures 2005; see: www.bbc.co.uk/radio4/reith2005/lectures.shtml (accessed 13 Aug 2005).

16   B Wynne, 'Public engagement or dialogue as a means of restoring public
        trust in science? Hitting the notes but missing the music', *Community
        Genetics* (2005, forthcoming).

17   J Wilsdon and R Willis, *See-through Science: Why public engagement needs to
        move upstream* (London: Demos, 2004); see chapter 2. This argument builds
        on a large body of social science research, for example: A Irwin and B Wynne
        (eds), *Misunderstanding Science? The public reconstruction of science and
        technology* (Cambridge: CUP, 1996).

18   HM Treasury/DTI/DfES, *Science and Innovation Investment Framework
        2004–2014*.

19   T Friedman, *The World is Flat* (London: Allen Lane, 2005).

20   'US slide in world share continues as European Union, Asia Pacific advance',
        ScienceWatch (Jul/Aug 2005); available at: www.sciencewatch.com/july-
        aug2005/sw_july-aug2005_page1.htm (accessed 13 Aug 2005).

21   T Branigan, 'Brown looks east to map UK's future', *Guardian*, 24 Feb 2005.

22   'Blair's failure', *Nature* 435 no 7039 (12 May 2005).

23   HM Treasury/DTI/DfES, *The Ten-year Science and Innovation Investment
        Framework Annual Report 2005* (London: HM Treasury, July 2005).

24   HM Treasury/DTI/DfES, *Lambert Review of Business–University
        Collaboration: Final report* (London: HM Treasury, Dec 2003).

25   S Rose, 'I'll show you mine if…', *Guardian*, 2 Jun 2005.

26   T Bentley, *Everyday Democracy: Why we get the politicians we deserve*
        (London: Demos, 2004).

27   G Orwell, 'Why I write', *Collected Essays Vol. 1* (London: Secker & Warburg,
        1968).

28   G Cook, *Genetically Modified Language* (London: Routledge, 2004).

29   Ibid.

30   T Blair, 'Common sense culture not compensation culture', Speech to the
        ippr, 26 May 2005.

31   D King, 'Improving science coverage in the media – invitation to a
        stakeholder communications workshop', letter to the authors, 6 Jul 2005.

32   We are grateful to Andy Stirling for this observation.

33   A Stirling, *Deliberate Futures: Precaution and progress in technology choice*
        (London: RICS, May 2005).

34   Y Ezrahi, *The Descent of Icarus: Science and the transformation of
        contemporary culture* (Cambridge MA: Harvard University Press, 1990).

35   HM Treasury/DTI/DfES, *Science and Innovation Investment Framework
        2004–2014*.

36   Ibid.

37   For more on this point, see J Wilsdon and R Willis, *See-through Science*.

38   B Bozeman and D Sarewitz, 'Public values and public failure in US science
        policy', *Science and Public Policy* 32 no 2 (April 2005).

39   MH Moore, *Creating Public Value: Strategic management in government*
        (Cambridge MA: Harvard University Press, 1995).

40   Ibid.

41   Ibid.
42   This account is taken from J Crabtree, 'The revolution that started in a library', *New Statesman*, 27 Sep 2004.
43   G Kelly and S Muers, *Creating Public Value: An analytical framework for public service reform* (London: Cabinet Office, 2002).
44   Strategy Unit, Note of a seminar on public value, 24 Sep 2002.
45   See www.bbc.co.uk/thefuture/bpv/prologue.shtml (accessed 13 Aug 2005).
46   J Holden, *Capturing Cultural Value* (London: Demos, 2004).
47   J Bend, *Public Value and e-Health* (London: ippr, 2004).
48   Crabtree, 'The revolution that started in a library'.
49   Quotes drawn from telephone interviews with Ted Freer and James Warner, July 2005.
50   Wilsdon and Willis, *See-through Science*.
51   Quoted in 'When disenchantment leads to disengagement', *Research Fortnight*, 26 Jan 2005.
52   'Going Public', *Nature* 431, no 7011 (21 Oct 2004).
53   The Royal Society and Royal Academy of Engineering, *Nanoscience and Nanotechnologies*.
54   D Taverne, 'Let's be sensible about public participation', *Nature* 432 (18 Nov 2004).
55   Wilsdon and Willis, *See-through Science*.
56   J Ziman, *Real Science: What it is and what it means* (Cambridge: CUP 2000).
57   H Brooks, 'The relationship between science and technology', *Research Policy* 23 (1994).
58   H Nowotny, 'Society in science: the next phase in an impetuous relationship', keynote speech at the Science and Society Forum, Brussels, 9–11 Mar 2005.
59   Interview with Mark Welland, 22 Jun 2005.
60   LE Kay, 'Problematizing basic research in molecular biology' in A Thackray (ed), *Private Science: Biotechnology and the rise of molecular sciences* (Philadelphia: University of Pennsylvania Press, 1998).
61   We are grateful to Andy Stirling for his thoughtful insights on this point.
62   R Jackson, F Barbagello and H Haste, 'Strengths of public dialogue on science-related issues', *Critical Review of International Social and Political Philosophy* 8, no 3 (Sep 2005).
63   Ibid.
64   Sir John Enderby, speaking at the launch of *See-through Science*, 1 Sep 2004.
65   Interview with Elizabeth Mitchell, 27 Jul 2005.
66   Comments at AGPI meeting, 13 Jul 2005.
67   European Science Social Forum Network, *Framework Programme 7: Towards a real partnership with society*, 2005; see: www.essfnetwork.org/campaigns.html (accessed 13 Aug 2005).
68   Interview with Claudia Neubauer, 22 Jul 2005.
69   The Royal Society and Royal Academy of Engineering, *Nanoscience and Nanotechnologies*.
70   HM Government, *The Government's Outline Programme for Public*

Engagement on Nanotechnologies, August 2005; see: www.ost.gov.uk/
policy/issues/programme12.pdf (accessed 13 Aug 2005).

71 M Welland, 'Now we're going public', *Guardian*, 19 May 2005.

72 Interview with Mark Welland.

73 Interview with Alexis Vlandas, 5 Jul 2005.

74 B Durodie, 'Limitations of public dialogue in science and the rise of new
    "experts"', *Critical Review of International Social and Political Philosophy* 6 no
    4 (2003).

75 A Irwin, *Citizen Science: A study of people, expertise and sustainable
    development* (London: Routledge, 1995).

76 J Higgins, 'Responsibility and science?'.

77 RK Merton, 'The normative structure of science' (1942) in N Storer (ed), *The
    Sociology of Science* (Chicago: University of Chicago Press, 1973).

78 For a critique of Merton's norms, see JM Mulkay, 'Norms and ideology in
    science', *Social Science Information* 15 no 4/5 (1975).

79 Sociologist Harry Collins has spent his career studying the replication of
    scientific experiments. See for example, H Collins, *Gravity's Shadow: The
    search for gravitational waves* (Chicago: University of Chicago Press, 2004).

80 DH Guston and D Sarewitz, 'Real-time technology assessment', *Technology in
    Society* 24 no 1 (2002).

81 www.royalsoc.ac.uk/page.asp?id=3180 (accessed 13 Aug 2005).

82 Quoted in S Boseley, 'From malaria to TB, top scientists get £245m to
    revolutionise world health', *Guardian*, 28 Jun 2005.

83 G Mulgan, 'Labour Britain' *Fabian Review* 117 no 2 (Summer 2005).

84 The Royal Society and Royal Academy of Engineering, *Nanoscience and
    Nanotechnologies*.

85 Discussed in chapter 5 of Wilsdon and Willis, *See-through Science*.

86 J Washburn, *University Inc: The corporate corruption of higher education* (New
    York: Basic Books, 2005).

87 M Bagioli, 'Galileo the emblem maker', *Isis* vol 81 (1990); M Norton Wise
    (ed), *The Values of Precision* (Princeton, NJ: Princeton University Press,
    1995).

88 A Ryan, 'Fading ivy', *Financial Times*, 4 Jun 2005.

89 C Langley, *Soldiers in the Laboratory: Military involvement in science and
    technology – and some alternatives* (Folkestone: Scientists for Global
    Responsibility, 2005).

90 A Nordmann, *Converging Technologies – Shaping the future of European
    societies* (Brussels: European Commission, 2004).

91 The Royal Society and Royal Academy of Engineering, *Nanoscience and
    Nanotechnologies*.

92 HM Government, *Response to the Royal Society and Royal Academy of
    Engineering Report*.

93 HM Treasury/DTI/DfES, *Science and Innovation Investment Framework
    2004–2014*.

94 For example, a US$3 million initiative on emerging nanotechnologies at the

Woodrow Wilson Center in Washington DC, and a major new programme at Arizona State University's Consortium for Science, Policy and Outcomes.

95 T Blair, 'Science matters'.

96 A Sen, *The Argumentative Indian: Writings on Indian history, culture and identity* (London: Allen Lane, 2005).

97 E Economy, '"China's environmental movement": testimony before the Congressional Executive Commission on China, Roundtable on Environmental NGOs in China' (Washington, DC: Council on Foreign Relations, 7 Feb 2005).

98 See for example, ME Porter and C van der Linde, 'Toward a new conception of the environment–competitiveness relationship', *Journal of Economic Perspectives* 9 no 4 (Fall 1995); P Hawken, AB Lovins and HL Lovins, *Natural Capitalism: The next industrial revolution* (London: Earthscan, 2000).

99 Demos has recently embarked on a project for the Foreign Office and other partners which will explore these and other dimensions of the new 'geography of science'. We hope this will contribute some helpful insights when the final report is published in October 2006. More details can be found at: www.demos.co.uk/projects/currentprojects/atlasproject/ (accessed 13 Aug 2005).

DEMOS – Licence to Publish

THE WORK (AS DEFINED BELOW) IS PROVIDED UNDER THE TERMS OF THIS LICENCE ("LICENCE"). THE WORK IS PROTECTED BY COPYRIGHT AND/OR OTHER APPLICABLE LAW. ANY USE OF THE WORK OTHER THAN AS AUTHORIZED UNDER THIS LICENCE IS PROHIBITED. BY EXERCISING ANY RIGHTS TO THE WORK PROVIDED HERE, YOU ACCEPT AND AGREE TO BE BOUND BY THE TERMS OF THIS LICENCE. DEMOS GRANTS YOU THE RIGHTS CONTAINED HERE IN CONSIDERATION OF YOUR ACCEPTANCE OF SUCH TERMS AND CONDITIONS.

1. **Definitions**
   a **"Collective Work"** means a work, such as a periodical issue, anthology or encyclopedia, in which the Work in its entirety in unmodified form, along with a number of other contributions, constituting separate and independent works in themselves, are assembled into a collective whole. A work that constitutes a Collective Work will not be considered a Derivative Work (as defined below) for the purposes of this Licence.
   b **"Derivative Work"** means a work based upon the Work or upon the Work and other pre-existing works, such as a musical arrangement, dramatization, fictionalization, motion picture version, sound recording, art reproduction, abridgment, condensation, or any other form in which the Work may be recast, transformed, or adapted, except that a work that constitutes a Collective Work or a translation from English into another language will not be considered a Derivative Work for the purpose of this Licence.
   c **"Licensor"** means the individual or entity that offers the Work under the terms of this Licence.
   d **"Original Author"** means the individual or entity who created the Work.
   e **"Work"** means the copyrightable work of authorship offered under the terms of this Licence.
   f **"You"** means an individual or entity exercising rights under this Licence who has not previously violated the terms of this Licence with respect to the Work, or who has received express permission from DEMOS to exercise rights under this Licence despite a previous violation.
2. **Fair Use Rights.** Nothing in this licence is intended to reduce, limit, or restrict any rights arising from fair use, first sale or other limitations on the exclusive rights of the copyright owner under copyright law or any other applicable laws.
3. **Licence Grant.** Subject to the terms and conditions of this Licence, Licensor hereby grants You a worldwide, royalty-free, non-exclusive, perpetual (for the duration of the applicable copyright) licence to exercise the rights in the Work as stated below:
   a to reproduce the Work, to incorporate the Work into one or more Collective Works, and to reproduce the Work as incorporated in the Collective Works;
   b to distribute copies or phonorecords of, display publicly, perform publicly, and perform publicly by means of a digital audio transmission the Work including as incorporated in Collective Works;
   The above rights may be exercised in all media and formats whether now known or hereafter devised. The above rights include the right to make such modifications as are technically necessary to exercise the rights in other media and formats. All rights not expressly granted by Licensor are hereby reserved.
4. **Restrictions.** The licence granted in Section 3 above is expressly made subject to and limited by the following restrictions:
   a You may distribute, publicly display, publicly perform, or publicly digitally perform the Work only under the terms of this Licence, and You must include a copy of, or the Uniform Resource Identifier for, this Licence with every copy or phonorecord of the Work You distribute, publicly display, publicly perform, or publicly digitally perform. You may not offer or impose any terms on the Work that alter or restrict the terms of this Licence or the recipients' exercise of the rights granted hereunder. You may not sublicence the Work. You must keep intact all notices that refer to this Licence and to the disclaimer of warranties. You may not distribute, publicly display, publicly perform, or publicly digitally perform the Work with any technological measures that control access or use of the Work in a manner inconsistent with the terms of this Licence Agreement. The above applies to the Work as incorporated in a Collective Work, but this does not require the Collective Work apart from the Work itself to be made subject to the terms of this Licence. If You create a Collective Work, upon notice from any Licencor You must, to the extent practicable, remove from the Collective Work any reference to such Licensor or the Original Author, as requested.
   b You may not exercise any of the rights granted to You in Section 3 above in any manner that is primarily intended for or directed toward commercial advantage or private monetary compensation. The exchange of the Work for other copyrighted works by means of digital file-

sharing or otherwise shall not be considered to be intended for or directed toward commercial advantage or private monetary compensation, provided there is no payment of any monetary compensation in connection with the exchange of copyrighted works.

c If you distribute, publicly display, publicly perform, or publicly digitally perform the Work or any Collective Works, You must keep intact all copyright notices for the Work and give the Original Author credit reasonable to the medium or means You are utilizing by conveying the name (or pseudonym if applicable) of the Original Author if supplied; the title of the Work if supplied. Such credit may be implemented in any reasonable manner; provided, however, that in the case of a Collective Work, at a minimum such credit will appear where any other comparable authorship credit appears and in a manner at least as prominent as such other comparable authorship credit.

5. Representations, Warranties and Disclaimer

a By offering the Work for public release under this Licence, Licensor represents and warrants that, to the best of Licensor's knowledge after reasonable inquiry:

    i Licensor has secured all rights in the Work necessary to grant the licence rights hereunder and to permit the lawful exercise of the rights granted hereunder without You having any obligation to pay any royalties, compulsory licence fees, residuals or any other payments;

    ii The Work does not infringe the copyright, trademark, publicity rights, common law rights or any other right of any third party or constitute defamation, invasion of privacy or other tortious injury to any third party.

b EXCEPT AS EXPRESSLY STATED IN THIS LICENCE OR OTHERWISE AGREED IN WRITING OR REQUIRED BY APPLICABLE LAW, THE WORK IS LICENCED ON AN "AS IS" BASIS, WITHOUT WARRANTIES OF ANY KIND, EITHER EXPRESS OR IMPLIED INCLUDING, WITHOUT LIMITATION, ANY WARRANTIES REGARDING THE CONTENTS OR ACCURACY OF THE WORK.

6. Limitation on Liability. EXCEPT TO THE EXTENT REQUIRED BY APPLICABLE LAW, AND EXCEPT FOR DAMAGES ARISING FROM LIABILITY TO A THIRD PARTY RESULTING FROM BREACH OF THE WARRANTIES IN SECTION 5, IN NO EVENT WILL LICENSOR BE LIABLE TO YOU ON ANY LEGAL THEORY FOR ANY SPECIAL, INCIDENTAL, CONSEQUENTIAL, PUNITIVE OR EXEMPLARY DAMAGES ARISING OUT OF THIS LICENCE OR THE USE OF THE WORK, EVEN IF LICENSOR HAS BEEN ADVISED OF THE POSSIBILITY OF SUCH DAMAGES.

7. Termination

a This Licence and the rights granted hereunder will terminate automatically upon any breach by You of the terms of this Licence. Individuals or entities who have received Collective Works from You under this Licence, however, will not have their licences terminated provided such individuals or entities remain in full compliance with those licences. Sections 1, 2, 5, 6, 7, and 8 will survive any termination of this Licence.

b Subject to the above terms and conditions, the licence granted here is perpetual (for the duration of the applicable copyright in the Work). Notwithstanding the above, Licensor reserves the right to release the Work under different licence terms or to stop distributing the Work at any time; provided, however that any such election will not serve to withdraw this Licence (or any other licence that has been, or is required to be, granted under the terms of this Licence), and this Licence will continue in full force and effect unless terminated as stated above.

8. Miscellaneous

a Each time You distribute or publicly digitally perform the Work or a Collective Work, DEMOS offers to the recipient a licence to the Work on the same terms and conditions as the licence granted to You under this Licence.

b If any provision of this Licence is invalid or unenforceable under applicable law, it shall not affect the validity or enforceability of the remainder of the terms of this Licence, and without further action by the parties to this agreement, such provision shall be reformed to the minimum extent necessary to make such provision valid and enforceable.

c No term or provision of this Licence shall be deemed waived and no breach consented to unless such waiver or consent shall be in writing and signed by the party to be charged with such waiver or consent.

d This Licence constitutes the entire agreement between the parties with respect to the Work licensed here. There are no understandings, agreements or representations with respect to the Work not specified here. Licensor shall not be bound by any additional provisions that may appear in any communication from You. This Licence may not be modified without the mutual written agreement of DEMOS and You.